IMAGES
of America

PORT OF TACOMA

This 1885 perspective map shows Tacoma and Commencement Bay. In 1873, the Northern Pacific (NP) Railroad had chosen Commencement Bay as the western terminus of its transcontinental railroad. The map highlights shipping activity in the bay and dock facilities along Tacoma's shoreline. Washington did not become a state until 1889, so it is labeled "W.T."—Washington Territory—on this map. (Library of Congress, 75696669.)

ON THE COVER: On Friday morning, March 25, 1921, a crowd of people welcomed the *Edmore*, the first ship to ever call at the Port of Tacoma. Railcars were shipside at Pier 1 with Tacoma-made lumber for the vessel. Longshore workers loaded the vessel in record time, and the ship set sail for Yokohama, Japan, the next day. (Port of Tacoma.)

Rod Koon

ISBN 978-1-4671-0984-0

Published by Arcadia Publishing
Charleston, South Carolina

Printed in the United States of America

Library of Congress Control Number: 2023931832

For all general information, please contact Arcadia Publishing:
Telephone 843-853-2070
Fax 843-853-0044
E-mail sales@arcadiapublishing.com
For customer service and orders:
Toll-Free 1-888-313-2665

Visit us on the Internet at www.arcadiapublishing.com

To all the people whose vision, ideas, dedication, and hard work have helped develop the Port of Tacoma over the last 100 years—and to all those who will help build and shape its future over the next 100 years

CONTENTS

Acknowledgments

From a historical perspective, I want to acknowledge that the Port of Tacoma and the Tideflats are situated on the shores of the Salish Sea and reside on the ancestral lands of the Puyallup tribe of Indians.

This book would not have been possible without the amazing help, guidance, and encouragement I received from many organizations and individuals.

I want to thank the staff at the Puget Sound Branch of the Washington State Archives, especially archivist Midori Okazaki. They helped me access many key port historical photographs and reference materials.

I am also very appreciative to these organizations and people for their help: the Northwest Room of the Tacoma Public Library (Spencer Bowman, Ilona Perry, and Anna Trammell), Tacoma Historical Society (Elizabeth Korsmo and Jessica Perry), and the Washington State Historical Society (Eileen Price).

Port of Tacoma staff members—current and former—also gave me valuable assistance: Jim Amador, Joe Barrentine, Leslie Barstow, Carol Bua, and Evette Mason. Diane Jordan provided invaluable help to me on my public records requests.

I also want to thank Bill Baarsma, Mike Jagielski, Eric Johnson, and Clare Petrich for reviewing various parts of the book and providing their valuable input.

I am also grateful to Dr. Ronald Magden, a talented historian and friend of the "working waterfront," who passed away in 2018. I had the honor of working with Ron on many port projects over the years. His two excellent books about the history of Tacoma's longshore were extremely helpful to me on this project.

And to my amazing wife, Tracey, I thank her for all the support and understanding she gave me while I worked on this book. I also want to thank my daughter Jessie for all her fine photography work on the book.

While this book captures many of the key highlights and milestones of the port's history, it includes some highlights and milestones of our region's history as well. I did that to put the port story into a larger context. I hope you will enjoy and appreciate that approach.

INTRODUCTION

Tacoma is a city whose early days were shaped by railroads, an abundance of natural resources, and a naturally deep, protected harbor.

So, it is not surprising that these same factors played major roles in the creation and development of the Port of Tacoma as well.

In the late 1800s, when the Northern Pacific Railroad came to Tacoma, it created a new era of growth and economic opportunities for the region. As an incentive for locating in Tacoma, the railroad was given valuable land along the city's shorelines. The railroad developed docks and shipping facilities there and had a virtual monopoly on who could use its facilities and at what price.

In the early 1900s, there was a growing movement among people who believed that public ownership of port facilities could break the railroad's waterfront monopoly and provide greater economic benefits to the public. Farmers and manufacturers around the state believed public ownership of ports would provide better access and lower rates for shipping their goods to market. Longshore workers saw public ports as a way to gain more work.

After many years of spirited discussion and debate on the issue—and railroad opposition—the Washington State Legislature passed the Port District Act in 1911. The act allowed voters of any county in Washington to create a port district encompassing all or part of the county.

While the initial attempt to create the Port of Tacoma failed in 1912, the second attempt was successful. On November 5, 1918, Pierce County voters approved the creation of the port by an overwhelming five-to-one margin. A year later, voters approved a port development plan and funds to get the port started.

To transform that development plan into reality, the port bought 240 acres of land in the Tideflats—a large marshy area of land and waterways located on Commencement Bay—not far from downtown Tacoma. The port embarked on developing shipping facilities that would create new jobs and opportunities for the residents of Pierce County—the true "stakeholders" of the port.

When the port opened Piers 1 and 2 in the 1920s, logs and lumber were two of the major cargoes handled there. Tacoma was known as the "Lumber Capital of America." For years, the port was also a major log-handling port, a cargo it handled well into this century.

In 1930, the port built a grain terminal, and a cold storage facility a year later. These two facilities helped fulfill promises that port proponents had made to farmers in rural Pierce County back in 1918. Locally produced fruit and produce could now be stored in the cold storage facility prior to shipment. In addition, wheat from eastern Washington could be moved by train to the port to be loaded onto ships.

In 1939, the Washington State Legislature passed a law that enabled ports to create Industrial Development Districts and levy a property tax for a limited time to provide funding for ports to foster economic development within their district. This law enabled the port to establish the Commencement Bay Industrial Development District in the Tideflats.

But the Great Depression and World War II delayed the port's industrial development efforts. When the war ended, more than 30,000 shipbuilding jobs in the Tideflats disappeared. That is when the port made key investments to develop new industrial lands and expand its waterways to attract much-needed jobs to the region.

Working with Pierce County, the City of Tacoma, and the Tacoma Chamber of Commerce, the port took the lead in developing a 300-acre tract of industrial land in the Tideflats known as the "Port Quadrangle." The area attracted many manufacturing and chemical companies, including Reichhold Chemical, Stauffer Chemical, and US Oil. The companies locating in the Quadrangle brought new jobs, cargoes, and commerce to the area.

Today, the era of major manufacturing and chemical companies in the Tideflats is largely over. But many of these companies left behind "legacy pollution," which the port and others are spending millions of dollars to clean up today.

In 1955, the port hired Tippetts, Abbett, McCarthy, Stratton (TAMS), a New York planning firm, to develop a plan for future port expansion, growth, and success. The plan called for $11 million of investments to extend the Hylebos and Blair waterways, which created additional space for berthing ships and about 1,500 acres of prime new industrial land in the Tideflats.

In August 1959, the port bought the 182-acre Tacoma Naval Station from the federal government. In 1968, the port created the Frederickson Industrial Development District and bought 510 acres of land at Frederickson, located 13 miles south of port terminals. Both of these strategic land purchases brought new companies, jobs, and economic activity to the region.

By the late 1970s, the era of containerized shipping revolution was well underway. These large metal boxes could carry anything from apples and frozen French fries to Almond Roca and hay. Containers also greatly increased the speed and efficiency of moving cargo.

Totem Ocean Trailer Express (TOTE), a major shipping line serving Alaska, moved from Seattle to Tacoma in 1976. This move helped the port earn the title of "Gateway to Alaska." In 1985, the top two container shipping lines in the world—Maersk Line and Sea-Land—both started calling at the port. The factors that attracted these and other lines to the port—available land, excellent intermodal rail connections, and a cooperative and productive longshore workforce—collectively became known as "The Tacoma Advantage."

In 1986, Tacoma's longshore workers, International Longshore and Warehouse Union Local 23, celebrated 100 years of organized labor on Tacoma's waterfront.

For many years, there was a spirited competition between the ports of Tacoma and Seattle for shipping line customers. For example, both TOTE and Sea-Land had previously called in Seattle. Other major container lines, including "K" Line and Evergreen Line—also moved from Seattle to Tacoma. Seattle's port also attracted some shipping lines that had been calling in Tacoma.

As other West Coast ports developed more container handling facilities and capacity, they also created more competition for the ports of Tacoma and Seattle. Eventually, both ports started to lose their market share of containerized cargo to other ports. In 2015, they decided to end their long-standing port rivalry and create The Northwest Seaport Alliance. This port partnership is primarily focused on marketing the port's container handling facilities and the benefits of shipping through the Puget Sound.

Today, in addition to containers, the port handles a diversity of other cargoes, such as automobiles, John Deere equipment, and other heavy industrial machinery. The port is also a strategic seaport for the military and handles a wide range of military equipment. Many of these cargoes are driven on and off ships equipped with loading ramps.

When voters created the Port of Tacoma in 1918, they hoped it could help create more jobs and economic vitality for Tacoma, Pierce County, and Washington state. Those hopes came true when the port developed piers, waterways, and land that brought new cargoes, new shipping lines, and new jobs to the region.

In the future, by making additional strategic investments, the port will continue to be a major economic engine, creating a world of opportunity for the people of Pierce County and Washington state.

One

The Foundations of a Great Port City

Many elements that helped shape the success of the Port of Tacoma were in place long before the public port was created: Commencement Bay was a sheltered, naturally deep harbor; Tacoma was the western terminus of Northern Pacific Railroad's transcontinental line; and the region had an abundance of natural resources that could be shipped to domestic and international markets. (Northwest Room, Tacoma Public Library, Richards Studio Richards-5.)

Northern Pacific Railroad's first train arrived in Tacoma on December 16, 1873. This milestone marked the completion of its transcontinental line. The railroad's presence in Tacoma brought new jobs, opportunities, and residents. The city grew from about 1,000 residents in 1870 to 30,000 in 1890. The railroad's headquarters (pictured) opened in 1888 in downtown Tacoma. (Northwest Room, Tacoma Public Library, Thomas H. Rutter Jogden-10.)

The Tacoma Hotel was built by the Tacoma Land Company, a railroad subsidiary. The luxurious hotel cost $267,000 to build and opened in 1887. Located in the heart of downtown Tacoma at 913 A Street, the hotel offered majestic views of Mount Rainier and Commencement Bay. It was destroyed by fire on October 17, 1935, and was never rebuilt. (Northwest Room, Tacoma Public Library, Richards Studio C117132-35.)

This photograph, taken around 1908, shows the Northern Pacific Railroad shops in South Tacoma. Known as the "South Tacoma Shops," the plant was composed of 36 separate brick buildings, covering 15 acres. Craftsmen from 20 different fields were hired to build, rebuild, and service anything that traveled on wheels for the Northern Pacific west of the Mississippi. (Northwest Room, Tacoma Public Library, Amzie D. Browning-080.)

A group of railroad workers poses in front of Locomotive No. 3013 on the track of the Northern Pacific's South Tacoma Shops. This photograph was taken around 1908. The railroad shops, which opened in 1891, had previously been located in downtown Tacoma. (Northwest Room, Tacoma Public Library, Amzie D. Browning-086.)

This 1888 photograph shows the three-masted clipper ship *Republic* unloading tea at the railroad wharves of the Northern Pacific Railroad. Tea was one of the many valuable and popular cargoes that moved across Tacoma's docks. Between 1888 and 1892, a total of 16 ships from Japan and China came to Commencement Bay bringing in 37 million pounds of tea. (Northwest Room, Tacoma Public Library, Thomas H. Rutter Jogden-08.)

The Foss Boathouse was the first boathouse in Tacoma. This photograph, taken around 1890, shows Andrew and Thea Foss and their daughter Lillian Foss. Andrew Foss built the structure. His wife, Thea, had bought a used rowboat for $5, and that marked the start of the Foss Launch and Tug Company. (Northwest Room, Tacoma Public Library, Marvin D. Boland-B4399.)

Taken around 1903, this photograph shows the crew of the *Buckingham* at a grain elevator along Tacoma's waterfront. The crew was under the command of William Roberts. It is believed that the woman in the photograph may have been a family member of one of the crew. (Northwest Room, Tacoma Public Library, Wilhelm Hester TPL-1097.)

The abundance of timber in Tacoma and the surrounding area led to the establishment of many lumber mills along the Tacoma waterfront. The region's lumber served as a magnet that brought hundreds of ships to Tacoma's piers to pick up the cargo. This photograph shows several ships loading lumber at a pier in 1909. (Washington State Historical Society, 1943.42.11832.)

Located on Tacoma's waterfront, the Hatch Mill was owned by Miles F. Hatch. It began operating on January 17, 1877, and closed in 1887. Other major lumber mills in the area at the time included the Hansen, Ackerman & Company Mill and Nicholas DeLin's mill. This photograph was taken around 1885. (Northwest Room, Tacoma Public Library, U.P. Hadley Paterson-01.)

In mid-June 1921, a large shipment of timber was loaded onto the *Genoa Maru* at the St. Paul & Tacoma Lumber Company. Two cranes were used to complete the loading. Located in the Tideflats, the mill was one of the largest in the area. It had been shipping lumber, especially Douglas fir, overseas for years. (Northwest Room, Tacoma Public Library, Marvin D. Boland-B4201.)

This photograph of a large log ready for shipment on a railroad flatbed was taken around 1907. The fir log was from the St. Paul & Tacoma Lumber Company, located in the Tideflats. The company had doubled the size of its Tacoma plant in 1901 and was producing vast amounts of lumber and shingles. (Northwest Room, Tacoma Public Library, Central News Company CNC-06.)

On February 19, 1924, workers at the St. Paul & Tacoma Lumber Company loaded a giant log onto a Chicago, Milwaukee & St. Paul railcar. The log was 72 inches in diameter and 40 feet long. It was sent on an exhibition tour in the Midwest and East to highlight the size and strength of Tacoma's forests and lumber industry. (Northwest Room, Tacoma Public Library, Marvin D. Boland-B9466.)

The Tidewater Mill, located on the east side of the Hylebos Waterway, was built in 1918 on six acres of soggy land in the Tideflats. It was one of the few West Coast mills that could cut timbers up to 130 feet long. With about 750 feet of dock space, it could load several vessels at a time. (Northwest Room, Tacoma Public Library, Marvin D. Boland-B4557.)

This 1921 photograph shows the Buffelen Lumber and Manufacturing Company, located at Lincoln Avenue and Taylor Way in the Tideflats. It was established in 1912 by John Buffelen primarily as a door manufacturer. In 1916, it also started making fir plywood. The company had offices in Tacoma and Minneapolis. (Northwest Room, Tacoma Public Library, Marvin D. Boland-B4952.)

A group posed in front of the "welcome log" at the 13th Annual Pacific Logging Congress Convention held in Tacoma in October 1922. This photograph is believed to have been taken at the St. Paul & Tacoma Lumber Company camp at Kapowsin. Delegates visited the camp on the last day of the convention. (Northwest Room, Tacoma Public Library, Marvin D. Boland-B6788.)

This photograph shows the three-masted ship *Eldorado* docked at the Northern Pacific Railroad's coal bunkers along Tacoma's waterfront around 1885. The railroad had earlier built a branch railroad up the Puyallup River Valley to help open the coalfields in Wilkeson, Carbonado, and Fairfax. In 1888, the Tacoma Coal and Coke Company opened the first commercial coke plant. (Northwest Room, Tacoma Public Library, C.E. and Hattie King-005.)

Four huge arches were built to span Pacific Avenue to welcome Pres. Benjamin Harrison to Tacoma on May 6, 1891. The arch shown in the photograph was built of coal taken from the mines in nearby Roslyn. The other arches featured additional leading regional products: wheat and flour, timber, and iron ore. The president's visit attracted thousands of people. (Northwest Room, Tacoma Public Library, TPL-2263.)

When this photograph was taken in 1920, Tacoma had four flour mills along its waterfront: Tacoma Grain Company, Sperry Flour Company, Puget Sound Flouring Mills, and Albers Brothers Milling Company. Together, the mills created the largest flour production west of Minneapolis and Kansas City. The Tacoma Grain Company, built in 1890, made Pyramid Flour (pictured). (Northwest Room, Tacoma Public Library, G34.1-111.)

By the early 1900s, Tacoma boasted first-class facilities for handling and shipping wheat. This photograph is labeled "Interior of the Largest Wheat Warehouse in the World." The wheat warehouses lined Tacoma's City Waterway (now Thea Foss Waterway). One portion of the warehouses was the Balfour Dock Building, which today is the home of the Foss Waterway Seaport. (Northwest Room, Tacoma Public Library, General Photograph Collection CNC-15.)

On March 13, 1920, a total of 400 tons of drifted snow flour were loaded onto the steamship *Edmore* docked at the Sperry Flour Company. On March 22, 1920, the *Tacoma Daily Ledger* ran a headline proclaiming, "Flour was the Greatest Tacoma Industry." A year later, the *Edmore* gained recognition as the first ship to call at the Port of Tacoma. (Northwest Room, Tacoma Public Library, Marvin D. Boland-B2711.)

In March 1920, longshore workers used dollies to load sacks of flour into the Japan-built freighter *Eastern Knight* at the Puget Sound Flouring Mills. After it was loaded with 10,000 tons of flour, the ship headed to the East Coast. In 1922, the Puget Sound Flouring Mills became part of the Sperry Flour Company. (Northwest Room, Tacoma Public Library, Marvin D. Boland-B2706.)

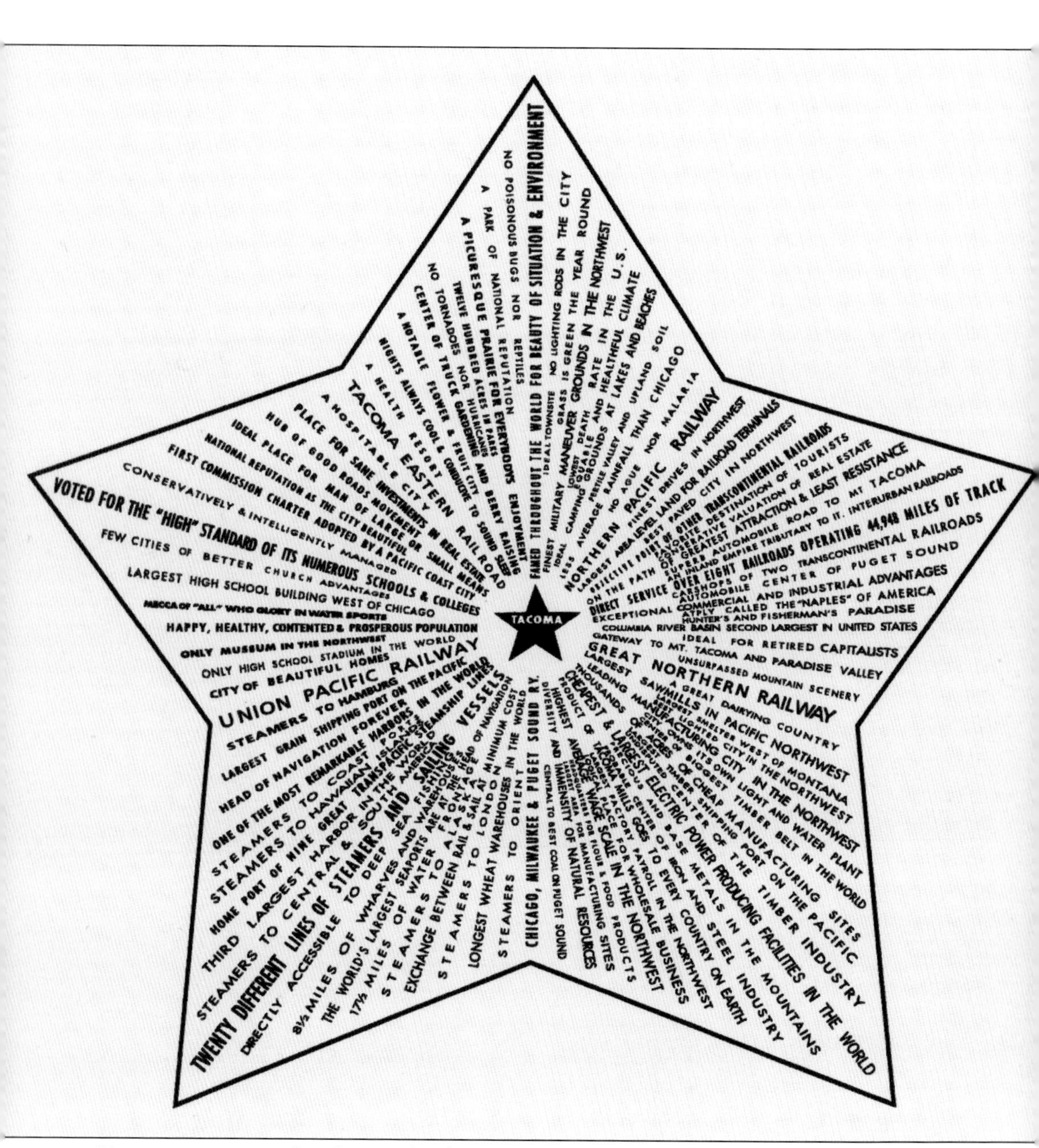

Tacoma promoter and real estate developer Allen C. Mason created this Star of Destiny design in 1910 to highlight Tacoma's many advantages. He used it in newspaper advertisements around the country to promote Tacoma. The advertisement listed more than 100 of the city's advantages, including its railroad and shipping connections. It also boasted that Tacoma had a "happy, healthy, contented and prosperous population." (Tacoma Historical Society.)

On May 1, 1911, more than 20,000 people attended the grand opening of the Union Station in downtown Tacoma. The *Tacoma Daily Ledger* called the occasion "the most brilliant public event in Tacoma's history." Designed by the Boston firm of Reed and Stem, the railroad station cost $750,000 to build. Tacoma's first rail station had opened in 1883. (Washington State Historical Society, 2011.0.211.)

The Tacoma Municipal Dock, opened in 1911, was the first publicly owned dock in Washington state. Tacoma mayor Angelo Fawcett championed a bond drive that voters approved to finance the project. The dock cost $270,850 to build. It was designed to handle passengers as well as cargo. The Tacoma Hotel rests on the bluff overlooking the dock building and Commencement Bay. (Northwest Room, Tacoma Public Library, TPL-1913.)

This April 29, 1918, photograph shows 10 ships under construction at the Foundation Company shipyards in the Tideflats. About 4,000 people worked at this site to build five-masted wooden schooners for the French government in World War I. The Foundation Company had five other yards around the country. The company was headed by Franklin Remington, who was part of the Remington firearms family. (Northwest Room, Tacoma Public Library, Foundation-5.)

On June 1, 1918, the *Gerbeviller* took a trial trip from Commencement Bay. Built in the Foundation shipyards, the 3,300-ton schooner had auxiliary steam engines and was 280 feet long. It was hailed as the first completed "all-Tacoma" wooden ship built since the United States entered World War I. (Northwest Room, Tacoma Public Library, Foundation-95.)

On December 14, 1920, the USS *Omaha* was launched at Todd's facilities in the Tacoma Tideflats. This was the 27th launching at the Todd's Tacoma yards. At the time, the 550-foot-long *Omaha* was the longest vessel ever to be launched in the Pacific Northwest. Todd built this scout cruiser for the Navy. (Northwest Room, Tacoma Public Library, Marvin D. Boland-B3345.)

On November 30, 1918, hundreds of shipyard employees and visitors gathered at Todd's yards for the launching of the *Jacona*. Todd Dry Dock & Construction Corporation, which had a variety of names over the years, was a major shipbuilder located between the Hylebos Waterway and the

Wapato Waterway (now Blair Waterway). (Northwest Room, Tacoma Public Library, Richards Studio C117132-41.)

The first steel bridge to span City Waterway (now Thea Foss Waterway) cost $90,000 to build and opened in 1894. Located at the foot of South Eleventh Street in downtown Tacoma, the bridge provided a key link between downtown Tacoma and the industries located in the Tideflats. The bridge was closed in 1911 and replaced by a vertical-lift bridge. (Northwest Room, Tacoma Public Library, Richards Studio C117132-3.)

The replacement for the original Eleventh Street Bridge opened in 1913. This steel truss lift bridge was built at the cost of $600,000. As shown in this 1927 photograph, the center section of the bridge could be raised to give clearance to large ships traveling to and from the upper sections of the waterway. (Northwest Room, Tacoma Public Library, Marvin D. Boland-B16995.)

This 1924 photograph shows a log dump on City Waterway. The logs were destined for use at the Wheeler-Osgood sawmill, shown in the background. Wheeler-Osgood eventually became the largest door manufacturer in the world. In 1939, their 27th million door was part of the Washington state exhibit at the New York World's Fair. The company employed 1,500 people at its peak. (Northwest Room, Tacoma Public Library, Marvin D. Boland-B9274.)

City Waterway was once a major center of industrial activity. This 1929 aerial photograph shows the Albers Brothers Milling Company at upper center and Wheeler-Osgood at right center. Other companies in the photograph include Northwest Woodenware Company, Independent Paper Stock Company, and Washington Gas & Electric Company. Union Station is in the upper left corner. (Northwest Room, Tacoma Public Library, Chapin Bowen TPL-672.)

This photograph, taken around 1907, shows logging activity near Tacoma. The steam donkey, shown in this picture, was a hoisting machine that was used to haul logs from the forests. These logs were sent to the nearly 40 sawmills, shingle mills, and woodworking plants that were operating in or around Tacoma in the early 1900s. (Northwest Room, Tacoma Public Library, Central News Company Central News Company CNC-04.)

Two

Port Creation and Early Developments

This 1901 photograph shows how much shipping activity was taking place on City Waterway adjacent to downtown Tacoma. Warehouses and docks for handling wheat and other commodities lined the waterfront. The Northern Pacific Railroad headquarters building, with its commanding view of shipping and rail activity along the waterfront, is in the upper left corner of the photograph. (Northwest Room, Tacoma Public Library, Arthur French Collection TPL-1344.)

After years of debate, the Washington State Legislature passed the Port District Act in 1911. The bill was designed to fight against the railroad's monopolies of waterfront lands in port cities. It also gave businesses, farmers, and others around the state greater access to shipping facilities. This photograph shows the original state capitol building. (Northwest Room, Tacoma Public Library, General Photograph Collection, the Coast Magazine Photograph WIL [F]-137.)

On September 5, 1911, King County citizens voted to create the Port of Seattle, the first public port in Washington state. On November 5, 1912, the ballot measure to create the Port of Tacoma failed by only 395 votes (10,581 to 10,186). Many people living in rural Pierce County believed that a public port would only benefit the citizens of Tacoma. (Seattle Municipal Archives, 316.)

Emma (Smith) DeVoe was one of the many leaders who promoted the second attempt to create the Port of Tacoma. In 1910, she had spearheaded the move that gave women the right to vote in Washington state. She had also worked with other states until women won the vote nationally in 1920. (Northwest Room, Tacoma Public Library, Richards Studio TPL-8717.)

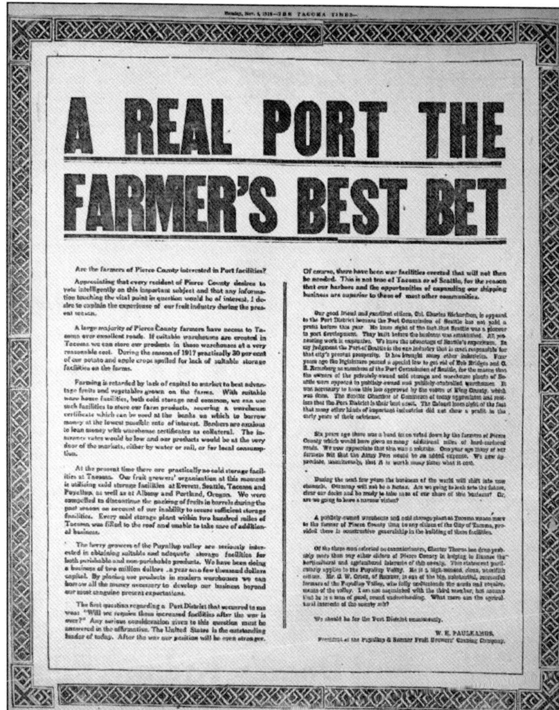

Monday, Nov. 4, 1918—THE TACOMA TIMES—

A REAL PORT THE FARMER'S BEST BET

Are the farmers of Pierce County interested in Port facilities?

Appreciating that every resident of Pierce County desires to vote intelligently on this important subject and that any information touching the vital point in question would be of interest, I desire to explain the experience of our fruit industry during the present season.

A large majority of Pierce County farmers have access to Tacoma over excellent roads. If suitable warehouses are erected in Tacoma we can store our products in these warehouses at a very reasonable cost. During the season of 1917 practically 30 per cent of our potato and apple crops spoiled for lack of suitable storage facilities on the farms.

Farming is retarded by lack of capital to market to best advantage fruits and vegetables grown on the farms. With suitable ware house facilities, both cold storage and common, we can use such facilities to store our farm products, securing a warehouse certificate which can be used at the banks on which to borrow money at the lowest possible rate of interest. Bankers are anxious to loan money with warehouse certificates as collateral. The insurance rates would be low and our products would be at the very door of the markets, either by water or rail, or for local consumption.

At the present time there are practically no cold storage facilities at Tacoma. Our fruit growers' organization at this moment is utilizing cold storage facilities at Everett, Seattle, Tacoma and Puyallup, as well as at Albany and Portland, Oregon. We were compelled to discontinue the packing of fruits in barrels during the past season on account of our inability to secure sufficient storage facilities. Every cold storage plant within two hundred miles of Tacoma was filled to the roof and unable to take care of additional business.

The berry growers of the Puyallup valley are seriously interested in obtaining suitable and adequate storage facilities for both perishable and non-perishable products. We have been doing a business of two million dollars a year on a few thousand dollars capital. By placing our products in modern warehouses we can borrow all the money necessary to develop our business beyond our most sanguine present expectations.

The first question regarding a Port District that occurred to me was: "Will we require these increased facilities after the war is over?" Any serious consideration given to this question must be answered in the affirmative. The United States is the outstanding leader of today. After the war our position will be even stronger.

Of course, there have been war facilities erected that will not then be needed. This is not true of Tacoma or of Seattle, for the reason that our harbors and the opportunities of expanding our shipping business are superior to those of most other communities.

Our good friend and [illegible] citizen, Col. Charles Richardson, is opposed to the Port District because the Port Commission of Seattle has not paid a profit before this year. He loses sight of the fact that Seattle was a pioneer in port development. They built before the business was established. All pioneering work is expensive. We have the advantage of Seattle's experience. In my judgment the Port of Seattle is the one industry that is most responsible for that city's present prosperity. It has brought many other industries. Four years ago the legislature passed a special law to get rid of Bob Bridges and C. E. Remsberg as members of the Port Commission of Seattle, for the reason that the owners of the privately-owned cold storage and warehouse plants of Seattle were opposed to publicly-owned and publicly-controlled warehouses. It was necessary to have this law approved by the voters of King County, which was done. The Seattle Chamber of Commerce of today appreciates and realizes that the Port District is their best asset. The Colonel loses sight of the fact that many other kinds of important industries did not show a profit in the early years of their existence.

Six years ago there was a bond issue voted down by the farmers of Pierce County which would have given us many additional miles of hard-surfaced roads. We now appreciate that this was a mistake. One year ago many of our farmers felt that the Army Post would be an added expense. We now appreciate, unanimously, that it is worth many times what it cost.

During the next few years the business of the world will shift into new channels. Germany will not be a factor. Are we going to look into the future, clear our docks and be ready to take care of our share of this business? Or, are we going to have a narrow vision?

A publicly-owned warehouse and cold storage plant at Tacoma means more to the farmer of Pierce County than to any citizen of the City of Tacoma, provided there is constructive generalship in the building of these facilities.

Of the three men selected as commissioners, Chester Thorne has done probably more than any other citizen of Pierce County in helping to finance the horticultural and agricultural interests of this county. This statement particularly applies to the Puyallup Valley. He is a high-minded, clean, unselfish citizen. Mr. C. W. Orton, of Sumner, is one of the big, substantial, successful farmers of the Puyallup Valley, who fully understands the needs and requirements of the valley. I am not acquainted with the third member, but assume that he is a man of good, sound understanding. What more can the agricultural interests of the county ask?

We should be for the Port District unanimously.

W. H. PAULHAMUS,
President of the Puyallup & Sumner Fruit Growers' Canning Company.

On November 4, 1918, full-page advertisements ran in local newspapers encouraging farmers to vote for the port. The advertisements were written by W.H. Paulhamus, the president of the Puyallup and Sumner Fruit Growers' Canning Company. On November 5, 1918, voters overwhelmingly approved the creation of the Port of Tacoma by a vote of 15,054 to 3,479. This advertisement ran in the *Tacoma Times* on November 4, 1918. (Northwest Room, Tacoma Public Library.)

Voters also elected the first three port commissioners in that same election. Edward Kloss received a total of 10,759 votes. Kloss came to Tacoma from California in 1906. At the time of the election, he had been involved in waterfront activities for more than 10 years. He also served as the business agent for the longshore union for many years. (Port of Tacoma.)

Charles Orton received 10,944 votes. He had a large fruit and dairy farm in Sumner and was the director of the Puyallup and Sumner Fruit Growers' Association. A huge garden party he held at his home in 1926 grew into what is now the Daffodil Festival. He was a strong advocate for a port cold storage facility that would benefit local farmers. (Port of Tacoma.)

Chester Thorne received the most votes in the election—11,412. Thorne came to Tacoma from California in 1906 and was very active in local business and financial circles. In 1920, he was the president of the board of directors of the National Bank of Tacoma and vice president of the Pacific Steamship Company. (Port of Tacoma.)

The first port commission meeting was held on November 19, 1918, in the National Bank of Tacoma Building (pictured on the right) on Pacific Avenue. In the meeting, Thorne was elected president of the commission, and Kloss was elected secretary. The commission also asked Thorne to arrange a meeting for the Tacoma port commissioners to call on the Seattle port commissioners. (Northwest Room, Tacoma Public Library, Marvin D. Boland-B15194.)

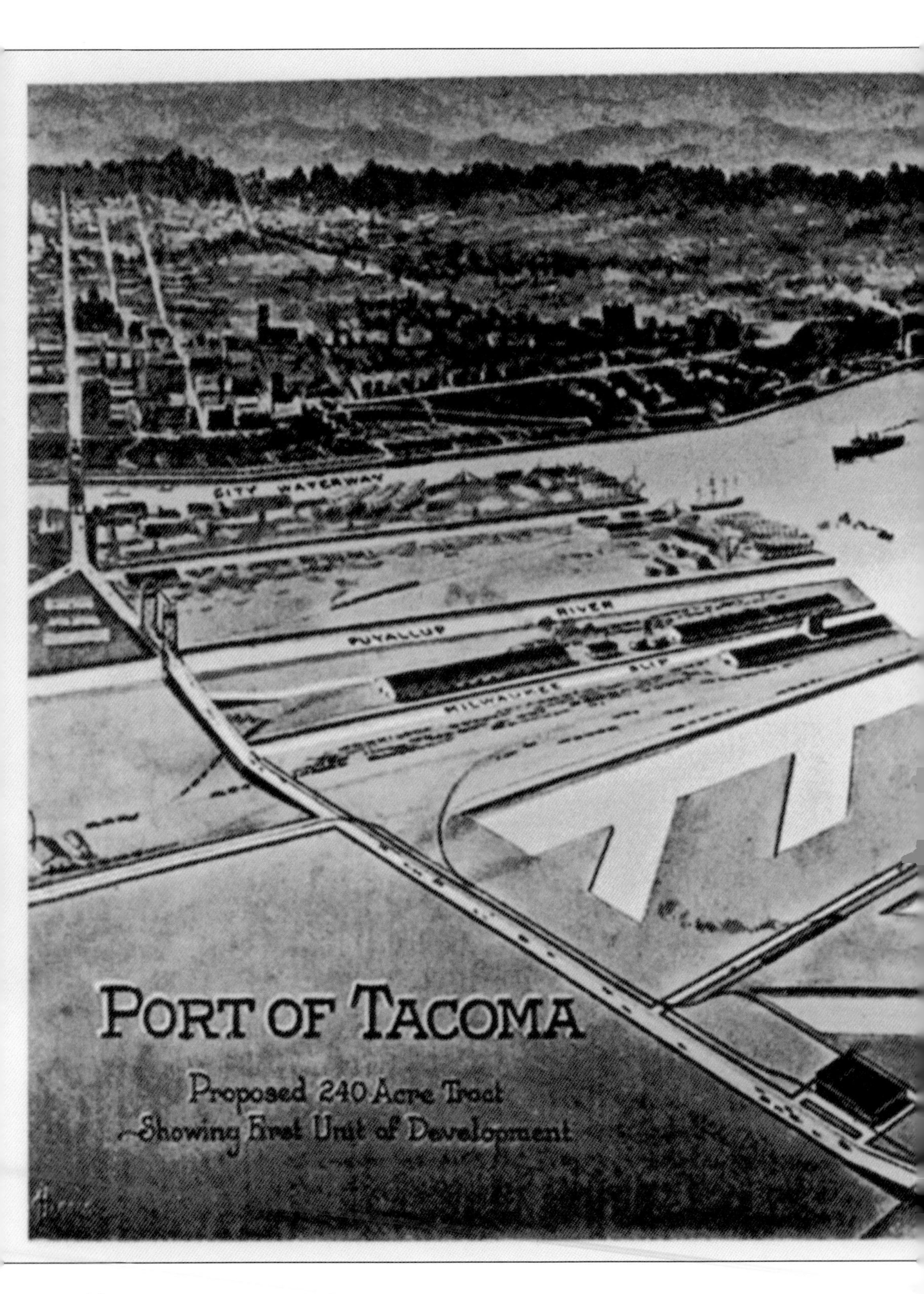

CITY WATERWAY
PUYALLUP RIVER
MILWAUKEE SLIP
PORT OF TACOMA
Proposed 240 Acre Tract
Showing First Unit of Development

Frank Walsh developed this plan for port development on 240 acres of land in the Tideflats. The port hired Walsh to update a previous port development plan by Virgil Bogue. On May 29, 1919, Walsh's plan was approved by a vote of 7,044 to 3,825. Pierce County voters also approved a $2.5 million bond issue to fund the development. (Port of Tacoma.)

Not long after the port development plan was approved, the port commission adopted this trademark insignia that reflected the plan. The commission stated that the drawing was "simply an outline of the units of the port development which will constitute the public dock facilities when all the piers and slips are completed." (Port of Tacoma.)

Even before its first pier was completed, the port was actively marketing the advantages its facilities would offer the shipping world. The port developed this trade show display to use at the National Foreign Trade Convention held in San Francisco on May 12–15, 1920. More than 20,000 delegates attended the convention. (Port of Tacoma.)

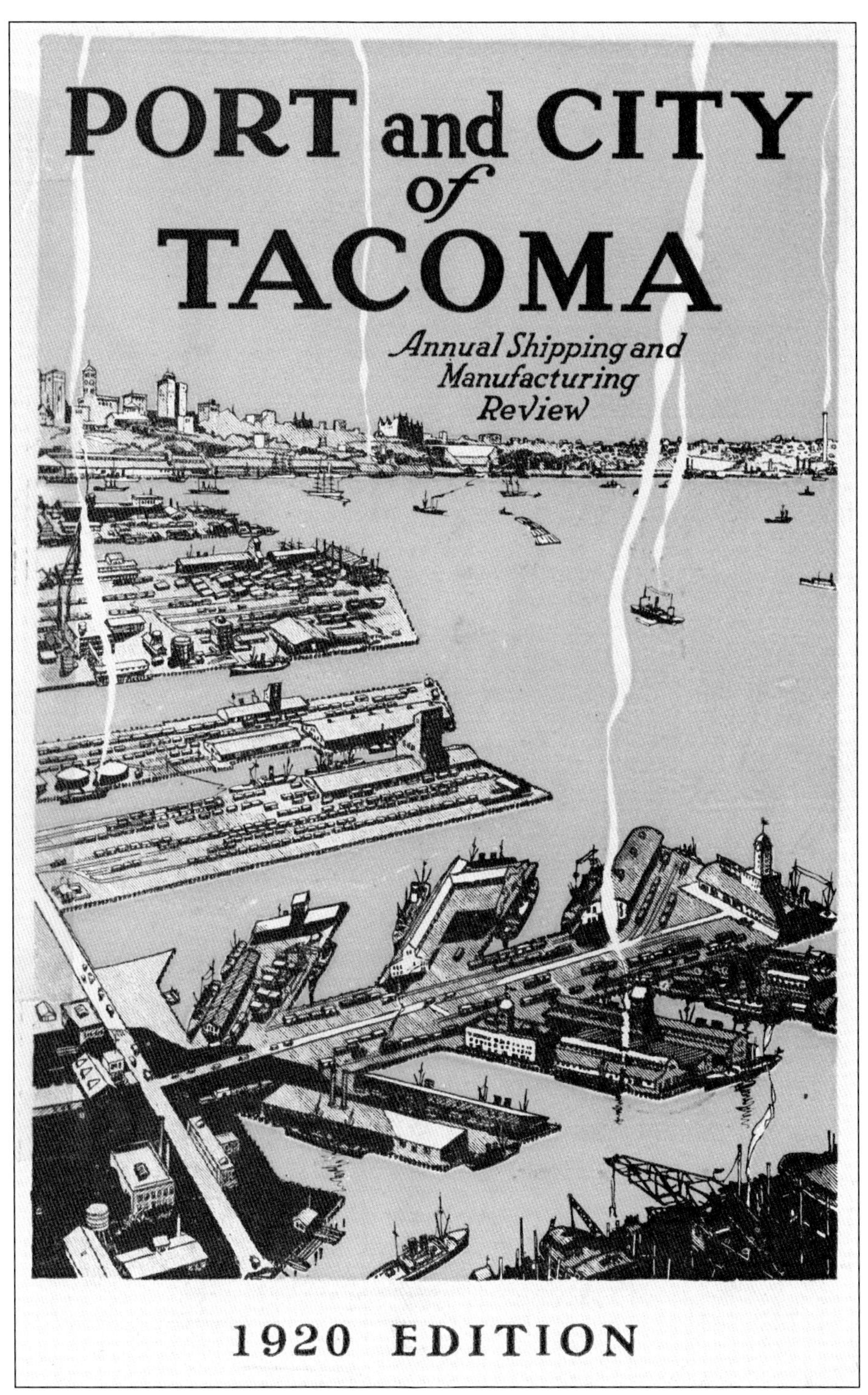

The port's planned facilities were depicted in the cover illustration on the port and City of Tacoma's 1920 annual shipping and manufacturing review. It also gave details of the port's development plans. The publication also highlighted Tacoma's status as a major lumber center, with more than $16 million of investments in lumber mills in Tacoma and Pierce County. (Port of Tacoma.)

Work on Pier 1 got underway in March 1920. This photograph shows a group of men getting set to spot the first pile that would be driven for Pier 1. The plans called for Pier 1 to be 800 feet long and 166 wide. Creosoted pilings would be used to support the pier structure. (Port of Tacoma.)

The Rutherford Company submitted the winning bid to the port commission to do the pile driving work for Pier 1. The company's bid amount was $169,599.88. The historic first pile was driven on March 15, 1920, by D.W. Rutherford. He was the head of the dredging company and also a Kiwanis club member. (Port of Tacoma.)

This photograph shows the "cutter" that was used in dredging and suction work on the Pier 1 project. The Tacoma Dredging Company dredged the waterway area to a depth of 35 feet at low tide. It also did bulkheading and riprap work for the project. The company's total bid was $404,980.50. The dredging was bid at 13.4¢ a cubic yard. (Port of Tacoma.)

This photograph, taken April 5, 1920, shows the Tacoma Dredging Company doing dredging work for Pier 1. The total construction time for Pier 1 was about one year. Originally designed to be 800 feet long, the pier was extended by an additional 300 feet in 1922. (Port of Tacoma.)

Once the wooden piles were driven for Pier 1, workers cut the pilings to their correct height to accommodate the pier decking. Today, concrete piles are used instead of wooden ones in pier construction. But in a very similar process, the concrete piles are sawed off to their proper height during the construction process. (Port of Tacoma.)

This photograph is believed to show the very first group tour of the Port of Tacoma's facilities. On August 4, 1920, members of the Commercial Club (part of the Tacoma Chamber of Commerce) came down to the port by boat to get a close-up look at the Pier 1 construction work that was underway. The group also had lunch on the pier. (Port of Tacoma.)

On March 21, 1921, the first train came onto Pier 1, in preparation for the arrival of the port's first commercial vessel. Two men are standing in front of the train, and some of the lumber the train brought in was from local mills. (Port of Tacoma.)

On Friday morning, March 25, 1921, port officials and longshore workers were on hand to welcome the *Edmore*—the first ship to call at the port. Railcars loaded with lumber were waiting shipside, having been delivered a few days earlier. Longshore workers loaded the vessel in record time. The ship left for Yokohama, Japan, the next day. (Port of Tacoma.)

RAILROAD CONNECTIONS

Freight shipped over **PORT OF TACOMA PIER ONE** may be routed via any of four great transcontinental railway systems and their connections, the Chicago, Milwaukee & St. Paul, Northern Pacific, Great Northern, and Union Pacific. Cars from these roads are switched alongside our sheds or to ships loading at our docks without extra charge to the shipper, thereby insuring fast and economical service.

PORT OF TACOMA
National Bank of Tacoma Bldg.
Tacoma, Washington

OCT 8 1921
OCT 15 1921
OCT 22 1921

From its earliest days, the port's rail connections have been a key to its success. This 1921 port advertisement highlights the port's rail connections and the fact it was served by four transcontinental railway systems: Chicago, Milwaukee & St. Paul; Northern Pacific; Great Northern; and Union Pacific. (Port of Tacoma.)

This photograph shows port engineers George Osgood (left) and Frank Walsh (right). Both men played key roles in developing and implementing the vision for the port's piers and related facilities. In February 1921, the commission picked Osgood to be the port's manager, in addition to his role as chief engineer. He served the port for more than 30 years, retiring in 1953. (Port of Tacoma.)

Pier 1 had enough open storage to accommodate 50 million board feet of lumber. While lumber was the leading cargo handled at Pier 1, the port handled other cargoes there as well—everything from cannons to large water pipes for the City of Tacoma. This photograph shows a crane and longshore workers maneuvering a boat called the *Idle Hour*. (Port of Tacoma.)

On April 30, 1925, the *Wheatland Montana* docked at Pier 1 to unload a shipment of mahogany logs from the Philippines. The logs were sent by rail to the nearby Buffelen Lumber Company plant and were used to make doors and panels. The vessel also unloaded 100 tons of peanuts. (Northwest Room, Tacoma Public Library, Marvin D. Boland-B12431.)

July 16, 1921 THE TRAFFIC WORLD 143

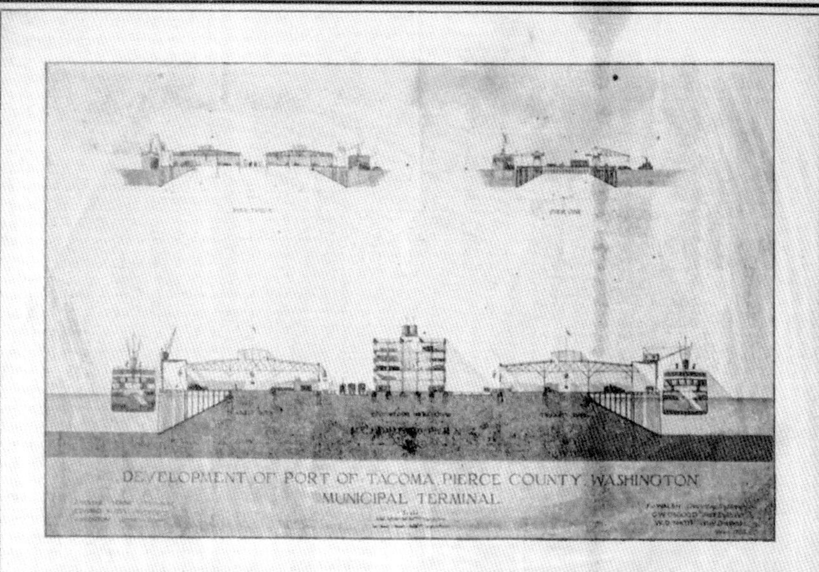

PORT OF TACOMA

Pierce County, State of Washington

MUNICIPAL RAIL & WATER TERMINAL

is being constructed according to typical pier sections as shown in the above cut. Situated on the waters of Commencement Bay, Puget Sound, it affords to Ocean Carriers one of the finest and safest harbors in the world. Special attention is being given to the installation of cargo handling equipment, looking toward the quick turn-around of ships and the fast dispatch of the cargo. For detailed information write to—

PORT OF TACOMA
National Bank of Tacoma B'ld'g.
Tacoma, Washington, U. S. A.

Soon after the port developed plans for Pier 2, a transit shed, and Pier 3, it started advertising those plans to the shipping industry. This advertisement appeared in the July 1921 issue of *Traffic World* magazine. The advertisement stated that these developments would help ensure "the quick turnaround of ships and the fast dispatch of cargo." (Port of Tacoma.)

Pier 2 was designed to be 674 feet wide and 1,200 feet long, with a transit shed nearby. The plan for the pier also included dockside rail to allow for a quick transfer of cargo between ship and rail. This photograph, taken July 20, 1922, shows two men and two horses working to grade Pier 2. (Port of Tacoma.)

In May 1922, the port commission awarded the contract to the Tacoma Dredging Company in the amount of $444,870.33 to build Shed No. 1 at Pier 2. The shed was designed to be 180 feet wide and 1,040 feet long. This photograph, taken September 23, 1922, shows the construction work that was well underway. (Port of Tacoma.)

This photograph, taken November 18, 1922, shows workers installing gunite wall panels at Shed No. 1. Gunite is a mixture of cement, sand, and water applied through a pressure hose, producing a dense hard layer of concrete. The gunite walls helped the port obtain a lower fire insurance rate for the shed. (Port of Tacoma.)

Dockside rail was an important component of Pier 1 and Pier 2. This 1925 photograph shows a port locomotive on Pier 2, with Shed No.1 in the background. Directly connecting to the port's tracks was the Municipal Beltline (now Tacoma Rail), owned by the City of Tacoma. It gave direct switching service to the transcontinental railroads serving the Tideflats. (Port of Tacoma.)

Two semi-portal cranes, known as Cranes 4 and 5, were designed to work with the monorail system in Shed No. 1 to move cargo more efficiently and cost effectively. This photograph, taken May 4, 1923, shows workers installing the two cranes at Pier 2. (Port of Tacoma.)

An innovative monorail system was another unique feature of Shed No. 1. The elevated monorail tracks ran from the center of the shed to the Pier 2 dock. This photograph, taken June 18, 1923, shows a worker installing a section of the elevated monorail track. (Port of Tacoma.)

This 1923 photograph shows longshore workers getting a sling load of lumber ready to be moved from inside Shed No. 1 to shipside at Pier 2 using the monorail crane system. The longshore driver in the monorail cab (also known as a telpher) was about 20 feet above the shed floor. (Port of Tacoma.)

Taken in 1926, this photograph shows longshore workers moving lumber from the shed to the ship using the monorail system and the semi-portal cranes. The telphers, which ran on a single circuit and in the same direction, were capable of reaching a speed of 700 feet per minute. (Port of Tacoma.)

The interior of Shed No. 1 had immense storage capacity and could accommodate a wide range of cargoes. "Jitney" vehicles were also used in the shed. Longshore workers drove the jitneys to move a variety of cargo to and from ships and around the piers. (Port of Tacoma.)

This photograph shows two ships berthed at Pier 2 with Shed No. 1 in the background. The equipment installed there was designed to help reduce a ship's time in port by one-third to one-half. The port's new facilities helped the total annual waterborne trade volumes moving through the Tacoma harbor grow from 2.7 million tons in 1919 to 6.4 million tons in 1929. (Port of Tacoma.)

These signs advertised the shipping lines and agents that were connecting the Port of Tacoma to the world. By 1929, the port had six services with Europe and Asia, four with South America, three with the Atlantic coast, and two within Puget Sound. The port also featured service with the Gulf Coast, the Mediterranean, and Australia. (Port of Tacoma.)

On July 5, 1923, Pres. Warren G. Harding arrived by train in Tacoma for a one-day visit. His visit included a reception in his honor at the Tacoma Hotel. This photograph shows him touring the US Veterans Hospital, where he visited disabled veterans. (Northwest Room, Tacoma Public Library, Marvin D. Boland G68.1-073.)

An estimated 25,000 people came to the Stadium Bowl in the rain to welcome President Harding and his wife to Tacoma. In his speech, the president highlighted the importance of the nation's ships and merchant marine. "There can be no dependable commerce without carriers," Harding said, "and there can be no eminence of American commerce without American carriers." (Northwest Room, Tacoma Public Library, Marvin D. Boland-B8287.)

After his speech, President Harding and his party came to the port and left Tacoma by ship from Pier 2. In honor of the visit, the port decorated the interior of Shed No. 1. A semi-portal crane at Pier 2 lowered the gangplank into position shipside that the president and his party used when they boarded their vessel. (Port of Tacoma.)

The USS *Henderson* docked at Pier 2 to pick up President Harding and his party, which included cabinet secretaries Herbert Hoover and Henry Wallace. Prior to leaving for Alaska, the ship sailed by Stadium Bowl to give the crowd one last chance to cheer on the president. He died 28 days later in San Francisco at the age of 57. (Port of Tacoma.)

This photograph, taken February 26, 1925, shows longshore workers securing an engine on a barge. The engine is believed to have been loaded onto the barge using a port crane at Pier 1. On the left in the distance, two vessels can be seen docked at Pier 2. (Port of Tacoma.)

Three

The Great Depression and World War II

As the Port of Tacoma entered its second decade of operation in the 1930s, the Great Depression brought unforeseen economic challenges to the port, the region, and the world. In 1930, US imports fell 13 percent and exports tumbled 22 percent. The port's tonnage dropped by 1.5 million tons in 1930. Despite these challenges, the port moved ahead on two major construction projects. (Port of Tacoma.)

On November 2, 1928, Pierce County voters approved a $500,000 bond issue to finance the construction of a grain elevator at the port. In 1929, the port spent about $12,000 on engineering work for the facility. This photograph shows the construction work that was well underway on the grain elevator in early 1930. (Port of Tacoma.)

When the grain facility opened in July 1930, its elevator had a capacity of 500,000 bushels, and its workhouse could handle 1.5 million bushels. The new facility enabled the port to export wheat to Europe. Called United Grain, the facility was designed to accommodate future expansion. (Port of Tacoma.)

This April 16, 1930, photograph shows workers pouring concrete for the foundation of the cold storage facility. The port built the facility to benefit local farmers and their crops. Back in 1917, it was estimated that about 30 percent of local potato and apple crops spoiled due to a lack of storage facilities for local farmers. (Port of Tacoma.)

When the cold storage facility became fully operational in 1931, it was handling about 275 barrels of berries a day during the summer. The *Tacoma Daily Ledger* newspaper proclaimed, "Not only has the port cold storage plant proven to be a good thing for Tacoma, but it has been a wonderful thing for the fruit and berries growers of the entire state." (Port of Tacoma.)

In late February 1931, at the depths of the Great Depression, about 1,000 men and women held a peaceful march on Tacoma City Hall to focus attention on immediate unemployment relief. Tacoma mayor M.G. Tennent met with 12 of the marchers to discuss in more detail their requests and the complexities of the situation. (Northwest Room, Tacoma Public Library, Chapin Bowen Collection TPL-1845.)

Not only did many people lose their jobs during the Great Depression, but some also lost their homes. As a result, a shantytown developed in the Tideflats, built by people who were in dire need of housing and shelter. The area became known as "Hollywood on the Tideflats." (Northwest Room, Tacoma Public Library, Richards Studio A7037-3.)

On June 15, 1933, as part of a nationwide tour, the historic USS *Constitution*, "Old Ironsides," was towed into Commencement Bay for a weeklong stay in Tacoma. More than 84,000 people toured the vessel while it was moored along Tacoma's shoreline. Launched in 1797, it is believed to be the world's oldest ship still afloat today. (Northwest Room at Tacoma Public Library, Richards Studio 639-1A.)

This 1939 aerial photograph shows the new Hylebos Bridge, just prior to its dedication. When both arms of the bascule bridge were in the upright position, its 150-foot opening enabled the safe passage of large ships. The Works Progress Administration gave a $170,000 grant to help fund the bridge, which cost a total of $400,000. (Northwest Room, Tacoma Public Library, Richards Studio D8358-8.)

Stadium High School majorettes posed with two Washington State Patrol officers on their motorcycles at the Hylebos Bridge dedication ceremony. The majorette team included (not listed in order) Bette Siegle, "Bert" Keely, Bettie Quinn, Mary Katherine Hager, Peggy Shaw, Billie Diederich, and Dollie White. (Northwest Room, Tacoma Public Library, Richards Studio D8379-8.)

This photograph shows members of the Hylebos Bridge celebration committee standing near the podium that was used in the bridge dedication ceremony on May 27, 1939. On the left is Joe Macek, the general chairman of the celebration. At the time of its opening, the bridge served around 3,500 residents in the area. (Northwest Room, Tacoma Public Library, Richards Studio D8379-1[B].)

Oscar Brown stands outside the Browns Point Lighthouse in this 1939 photograph. He had been the lighthouse keeper there since 1903, when the original wooden lighthouse structure was built. In 1934, the US Coast Guard built this new white concrete structure to replace the old one. Today, the lighthouse still helps guide ships safely in and out of Commencement Bay. (Northwest Room, Tacoma Public Library, Richards Studio D8456-9A.)

Once the new Browns Point Lighthouse was built, the fog bell used in the original lighthouse was no longer needed. Having been replaced by a modern automatic fog signal, the bell was donated to the College of Puget Sound (now University of Puget Sound). This photograph shows incoming freshmen around the bell during orientation week in September 1939. (Northwest Room, Tacoma Public Library, Richards Studio D8886-5A.)

This 1937 photograph shows a motion picture projectionist, an employee of the Works Progress Administration (WPA), at the Tacoma Armory. Adults and children were in the movie's audience. The WPA, created in May 1935, was a New Deal program designed to create jobs and income for the unemployed during the Great Depression. The program ended in 1943. (Northwest Room, Tacoma Public Library, Richards Studio A6007-1.)

For many years, the port housed a fishing fleet, which was moored at the Port Commission Docks near Pier 2. In 1934, Tacoma was the home port to about 45 fishing vessels. In the 1990s, the fleet was relocated to City Waterway (now Thea Foss Waterway) so the port could redevelop the marina area into a major container terminal. (Northwest Room, Tacoma Public Library, Richards Studio 872-3.)

Originally destined for Vladivostok, the American vessel *Wildwood* and crew (pictured) turned around in the mid-Pacific in April 1940 and returned to Tacoma and unloaded $4.5 million of cargo. It was feared the ship would run into a British blockade and its cargo would be confiscated. This event marked the extension of the European war into the Pacific trade routes. (Northwest Room, Tacoma Public Library, Richards Studio A9679-1.)

Tacoma's first fireboat (pictured) was launched in 1929. The Coastline Shipbuilding Company built the vessel at the cost of $148,000. For 53 years, the fireboat patrolled Tacoma's shoreline doing everything from fighting fires to handling water rescues. No longer in service, the fireboat is out of the water and now displayed on land at Marine Park on Tacoma's waterfront. (Northwest Room, Tacoma Public Library, Richards Studio D14468-22.)

This 1936 photograph shows William Pitcher (front left center), the Work Progress Administration foreman for the Fort Nisqually reconstruction project at Tacoma's Point Defiance Park, receiving a gold watch from G.A. Baker, the woods foreman on the project. The watch was a gift from the workers at the fort. The fort was built as a fur trading post in 1833 by the Hudson's Bay Company and was originally located in what is now DuPont, Washington. The fort was the first globally connected settlement on Puget Sound. (Northwest Room, Tacoma Public Library, Richard Studio T76-1.)

Taken January 13, 1940, this aerial view of port facilities shows two white military vessels berthed at Pier 2. A total of three ships were used to transport more than 7,500 soldiers stationed at nearby Fort Lewis to California for training. Across from the military vessels are two cargo ships berthed at Pier 1. (Northwest Room, Tacoma Public Library, Richards Studio D9334-6.)

The USS *Leonard Wood*, shown berthed at Pier 2, was one of the vessels that transported troops to California to participate in the largest "war games" ever held in US history. Two vessels from Olympia also transported troops to the war games. (Northwest Room, Tacoma Public Library, Richards Studio D9334-15.)

The USS *Republic* was the first ship to transport troops for the California war games. A total of 1,800 troops boarded the vessel at Pier 2 on January 3, 1940. During the war games, the Navy taught the troops how to load, lower away, row, and land a fleet of small boats through the California breakers. (Northwest Room, Tacoma Public Library, Richards Studio D9275-1.)

Henry Foss (far left), of the Foss Launch and Tug Company, was a featured speaker at the world premiere of the movie *Tugboat Annie Sails Again* at the Roxy Theater (now Pantages) on October 18, 1940. The movie featured Ronald Reagan as well as Foss tugs. Years later, Henry Foss served as a Port of Tacoma commissioner (1951–1952). (Northwest Room, Tacoma Public Library, Richards Studio D10341-30.)

This April 1943 photograph shows US Coast Guardsmen and firefighters from the Tacoma Fire Department burning down old, abandoned shacks to clean up the waterfront. The shacks were burned with the permission of the property owners. The effort was designed to help prevent fires in the Tideflats in case of an enemy attack. (Northwest Room, Tacoma Public Library, Richards Studio D14419-6.)

On April 17, 1941, the 205th Coast Artillery (antiaircraft) unit set up machine gun stations in the Tideflats to protect Tacoma from a mock "attack" by planes. Powerful searchlights were turned on at various locations at night to locate planes flying through Tacoma's skies. This camouflaged foxhole was set near the Hooker Chemical Company plant, which opened in 1929. (Northwest Room, Tacoma Public Library, Richards Studio D11147-5.)

President Roosevelt announced a nationwide scrap rubber collection campaign that ran in June 1942. Rubber recycling was necessary because most of America's crude rubber supply was cut off by Japan due to the war. Members of the Tacoma fishing fleet, located at the port, collected about 800 pounds for the drive, mostly old tires they used as bumpers. (Northwest Room, Tacoma Public Library, Richards Studio D13016-1.)

On August 4, 1942, Bing Crosby arrived by train at Tacoma's Union Station to support wartime bond sales. He performed with a USO troop at Fort Lewis and headlined a show at the Liberty Center along with Phil Silvers. Bing was born Harry Lillis Crosby in Tacoma on May 3, 1903, and lived at 1112 North J Street. (Northwest Room, Tacoma Public Library, Richards Studio D13217-1.)

On January 2, 1943, a gilded pick was used at the groundbreaking ceremony for the $5 million Seattle-Tacoma Airport. The ports of Tacoma and Seattle had worked together to make the airport a reality. Dignitaries at the event included Tacoma mayor Harry Cain; George Osgood, Port of Tacoma; and Port of Tacoma commissioners Charles Orton and Fred Marvin. (Northwest Room, Tacoma Public Library, Richards Studio D13938-3.)

This photograph shows employees of the Henry Mill and Timber Company, located at 3001 North Starr Street in Tacoma. During World War II, the mill had military contracts to build everything from cargo barges to prefabricated massive blimp hangars. The large cranes in the background were used to unload timber and load finished products onto railroad cars and ships. (Northwest Room, Tacoma Public Library, Richards Studio A16272-12.)

On August 15, 1945, the Boy Scouts served as the color guard at the Surrender Day Parade in front of troops parading through the business district of downtown Tacoma along Pacific Avenue. The parade included soldiers from Fort Lewis and McChord Field, Navy personnel, and veterans. Thousands of people attended the celebration marking the end of World War II. (Northwest Room, Tacoma Public Library, Richards Studio D20009-24.)

The USS *Admiral Eberle* was the first troop transport ship to dock at the Port of Tacoma since 1940. It arrived at Pier 2 on September 25, 1945. The vessel brought 134 officers and 4,369 enlisted men back from the war in the Pacific. (Northwest Room, Tacoma Public Library, Richards Studio D20371-6.)

Acting Tacoma mayor Val Fawcett (far right) greeted veterans returning on the transport USS *Admiral Coontz* on October 10, 1945. From left to right are Sgts. Robert Kusek, Ole Leland, and Robert Breen, Lt. John Watkins, and Capt. Kenneth Peterson. The ship brought 4,609 troops to the port from Okinawa. The troops traveled in convoys to nearby Fort Lewis for processing. (Northwest Room, Tacoma Public Library, Richards Studio D20490-6.)

In January 1946, a group of Red Cross volunteers had baskets of doughnuts and milk to share with servicemen returning on troop ships at the port. Due to a shortage of troop trains in the Pacific Northwest and a lack of local housing, thousands of soldiers had to temporarily stay aboard ships in the port. (Northwest Room, Tacoma Public Library, Richards Studio D21209-2.)

During World War II, women in Tacoma filled many of the jobs that had been traditionally occupied by men. This unidentified woman, holding a piece of wood against a planer, worked in one of Tacoma's many lumber mills. In 1942, there were 16 furniture manufacturing companies in Tacoma and 11 sawmills. (Northwest Room, Tacoma Public Library, Richards Studio D15730-3.)

Four

Shipbuilding in World War II

Throughout World War II, shipbuilding activities dominated the Tideflats and the Tacoma area. The Seattle-Tacoma Shipbuilding Corporation (later Todd-Pacific Shipyards) employed about 33,000 men and women at its wartime peak. The company was Tacoma's largest employer. This photograph, taken January 31, 1940, shows work underway to prepare shipbuilding facilities to meet the demands of the impending war. (Northwest Room, Tacoma Public Library, Richards Studio D9368-7.)

On August 1, 1940, hundreds of people watched the *Cape Alva* slide down the ways at the Seattle-Tacoma Shipbuilding Corporation plant. It was the first major motor ship to be built in Tacoma in 17 years. It was built in near record time, just four months after its keel was laid. The shipyard was located at 100 Alexander Avenue. (Northwest Room, Tacoma Public Library, Richards Studio D10090-25.)

Two women and a child sit on a blanket at the Seattle-Tacoma Shipbuilding Corporation company picnic, which was held August 4, 1940, at Five Mile Lake in south King County. The women were wearing huge sombreros to protect them from the sun. (Northwest Room, Tacoma Public Library, Richards Studio D10099-1.)

During the war, training programs were created to meet the demand for skilled shipyard workers. In this 1940 photograph, an instructor shows a student how to use a welding machine at Hawthorne School. The instructors were master welders on loan from the shipyards. The school district was reimbursed for the training costs by the national defense training program. (Northwest Room, Tacoma Public Library, Richards Studio D10545-1.)

On September 28, 1940, employees of the Seattle-Tacoma Shipbuilding Corporation posed for a team photograph prior to the launch of their second vessel, the *Cape Flattery*. It was built for the US Maritime Commission. The ship was named for the northernmost point in the contiguous United States, where the Strait of Juan de Fuca joins the Pacific Ocean. (Northwest Room, Tacoma Public Library, Richards Studio D10289-33.)

This aerial photograph, taken July 31, 1941, shows construction work at the Seattle-Tacoma Shipbuilding Corporation. Expansion was underway to meet the demands of new military contracts. Under construction were five additional new ways, more dock space, and expanded fabrication facilities. A total of 74 ships—ranging from escort aircraft carriers to cargo ships— were built here during World War II. (Northwest Room, Tacoma Public Library, Richards Studio D11693-13.)

On February 28, 1941, five women posed in front of the vessel *Cape Alava*, which had been launched months earlier as the *Cape Alva.* The ship was 416 feet long, 60 feet wide, and cost about $2 million to build. Three of the women in the photograph are believed to be Ella Wise, Lois Bergery, and Delphine Stewart. (Northwest Room, Tacoma Public Library, Richards Studio D10876-18.)

During World War II, many Tacoma shipyards received government contracts to build vessels to support the country's war effort. This 1944 aerial photograph shows the Western Boat Building Company, which specialized in the mass production of wooden ships for the Army that were used to transport passengers and supplies during the war. Martin Petrich Sr. started the company in 1916. (Northwest Room, Tacoma Public Library, Richards Studio D18088-3.)

This 1941 photograph shows a covered steel barge docked at J.M. Martinac Shipbuilders. The vessel, which was 110 feet long and 34 feet wide, was ordered by the Navy and fabricated by Birchfield Boilers. Martinac also built 12 wooden-hulled minesweepers for the Navy during the war. The company was located on City Waterway (now Thea Foss Waterway). (Northwest Room, Tacoma Public Library, Richards Studio D10822-6.)

The thousands of shipyard workers who came to Tacoma during the war created a dramatic housing shortage. This led to the development of the largest residential housing project ever attempted in Tacoma. Called Salishan, the project included 2,000 housing units. The Tacoma Housing Authority managed Salishan for the federal government. (Northwest Room, Tacoma Public Library, Richards Studio D14053-3.)

The Salishan development was located on 188 acres of land in southeast Tacoma. This photograph shows workers unloading sheets of plywood from a truck bed through an elevated doorway. The initial Salishan plan called for 1,600 of the housing units to be permanent and 400 units to be temporary. (Northwest Room, Tacoma Public Library, Richards Studio D14053-6.)

On Saturday, May 1, 1943, the first 10 families moved into their new housing units in Salishan. Pictured outside their new home is the Ralph J. Yorges family: Mr. and Mrs. Yorges, their baby Ralph, 10-year-old daughter Janice, and their dog. The family was originally from Palouse, Washington. (Northwest Room, Tacoma Public Library, Richards Studio D14451-4.)

Housing was not the only thing in short supply at Salishan—phones were, too. To save precious materials, the War Production Board only allowed Salishan 12 outdoor phone booths. This November 1943 photograph shows Murray Burgess using a phone. Waiting in line for their turn are, from left to right, Roy Ehlis, Marie Escarga, Mary Escarga, Andrew Ehlis, and R.T. Armstrong. (Northwest Room, Tacoma Public Library, Richards Studio D16374-4.)

By June 1942, more than 20,000 people were working the shipyards in the Tideflats—19,000 in ship production and another 1,000 in office jobs. The shipyard also had 130 people working in its own private, uniformed security force. At the time, the shipyard's security force was almost the same size as Tacoma's police force. (Northwest Room, Tacoma Public Library, Richards Studio D12808-4.)

The huge number of workers coming in and out of the shipyards every day created a major challenge for the Tideflats. Traffic jams were common at shift changes since most workers drove their own vehicles to and from work. To address the issue, a fleet of buses was brought into service to carry workers to and from the shipyards. (Northwest Room, Tacoma Public Library, Richards Studio D13359-2.)

On September 27, 1941, preparations were underway to launch the US Army transport *Frederick Funston* (pictured on the right) at the Seattle-Tacoma Shipbuilding Corporation's yards in Tacoma. The $3 million vessel was named in honor of the late Maj. Gen. Frederick Funston. Dignitaries at the celebratory launch event included Washington governor Arthur Langlie and Tacoma mayor Harry Cain. (Northwest Room, Tacoma Public Library, Richards Studio D11948-86.)

On May 4, 1943, Prince Alexander, the Earl of Athlone, and his wife, Princess Alice, visited the Seattle-Tacoma Shipbuilding facilities in Tacoma. Princess Alice (center) was presented with a corsage by shipyard welders Crystal Fender and Gladys Price and burner Frances Miller. The princess was a granddaughter of Queen Victoria and honorary commandant of the Women's Royal Canadian Naval Service. (Northwest Room, Tacoma Public Library, Richards Studio EW-429.)

During their shipyard visit, Princess Alice (third from left) and her husband, the Earl of Athlone, (second from left) viewed a scale model of the shipyard facilities and surrounding waterways. The Hylebos Waterway is displayed in the model near the group. The Earl of Athlone was governor-general of Canada and an uncle of King George VI. (Northwest Room, Tacoma Public Library, Richards Studio EW-422.)

On Mother's Day 1943, mothers who worked in defense plants and had sons serving in the military were honored. Esther Baker, who worked as a welder in the Tacoma shipyards, is pictured here. Her son Rawlin Charles Baker had also been a welder in the shipyards before joining the US Navy. (Northwest Room, Tacoma Public Library, Richards Studio D14496-3.)

In May 1943, Hazel Fuhrman was doing her part to support the war effort by working as a shipfitter's helper. Her 18-year-old son was serving in the US Navy. This photograph was published in the *Tacoma Times* on May 8, 1943, for the nation's second wartime Mother's Day. (Northwest Room, Tacoma Public Library, Richards Studio D14496-6.)

Mrs. C.E. Taylor, holding roses, posed with a group of unidentified women during the July 15, 1943, launching of the USS *Sunset*. She was the sponsor for the vessel, which was launched at the Tacoma yard of the Seattle-Tacoma Shipbuilding Corporation (later Todd-Pacific Shipyards Inc.). Mrs. Taylor was the wife of Comdr. C.E. Taylor, ordnance officer, 13th Naval District. (Northwest Room, Tacoma Public Library, Richards Studio 43-25.)

The USS *Sunset* was the 43rd ship launched by the Seattle-Tacoma Shipbuilding Corporation in its Tacoma yard. Although built for the US Navy, it was transferred to the British Royal Navy in November 1943 and commissioned as the HMS *Thane*. The vessel operated as a ferry and convoy transport-escort in the North Atlantic. (Northwest Room, Tacoma Public Library, Richards Studio 43-20.)

This photograph, taken December 18, 1942, shows hundreds of people gathered at the Tacoma Boat Building to see the company receive an "E" Award from the Army and Navy. The "E" Award was given to companies for excellence in wartime production. That same day, the company also launched its newest minesweeper. The plant was located at 2124 East D Street. (Northwest Room, Tacoma Public Library, Richards Studio D13882-6.)

Coast Guard lieutenant commander Jack Dempsey, a former heavyweight champion of the world, came to Tacoma on June 24, 1944, to help promote a war bond drive. He signed autographs for hundreds of fans during his visit, which included stops at Todd-Pacific Shipyards, McChord, Fort Lewis, and Point Defiance. (Northwest Room, Tacoma Public Library, Richards Studio D17858-6.)

On June 8, 1944, more than 15,000 spectators came to McChord Field to view the parade that kicked off the Fifth War Loan Campaign. Dignitaries in this photograph include bank executive Reno Odlin, (second from left), Washington governor Monrad Wallgren (far right), and two military officers who participated in the events. A military orchestra can be seen in the background. (Northwest Room, Tacoma Public Library, Richards Studio D17716-9.)

On July 15, 1944, thousands of residents gathered on Broadway in downtown Tacoma to see the Fifth War Loan Campaign Parade. The Todd-Pacific Shipyards float, one of many floats in the parade, featured a model ship on top. The parade was held to boost the sale of bonds during the loan drive and in honor of Infantry Day. (Northwest Room, Tacoma Public Library, Richards Studio D17771-2.)

Five

Postwar Economic Recovery

During World War II, the US Army took over most of the port's facilities to handle war cargo—everything from sacked flour for troops in Hawaii to blankets for soldiers in Alaska. By the war's end, West Coast port trade had dropped 90 percent. The port's postwar challenge was to get more cargo moving through Piers 1 and 2 (pictured). (Port of Tacoma.)

This 1948 aerial photograph, showing thousands of logs boomed together on port waterways, highlights the fact that logs were still one of the leading cargoes moving through the port. More than a dozen flattop carriers used in World War II can be seen in the distance. (Northwest Room, Tacoma Public Library, Richards Studio D35790-5.)

This 1947 photograph shows lumber from the Cheney Lumber Company on Pier 1. Ben Cheney originally founded the company to make railroad ties. He later redeveloped the 12-foot stud to make the more economical 8-foot stud, which helped revolutionize the home building industry. At the time, the port's facilities in this area were collectively called "Port Piers." (Northwest Room, Tacoma Public Library, Richards Studio D27484-7.)

In 1951, the port added 16 new elevators to United Grain, increasing its capacity by 500,000 bushels. The facility had a two million–bushel capacity by 1954 and could receive up to 100 cars of grain a day. The Archer-Daniels-Midland Company was leasing the facility. (Northwest Room, Tacoma Public Library, Richards Studio A63233-1.)

The *Gretna Victory* arrived at the Baker Dock in Tacoma on January 30, 1948, to pick up relief supplies and clothing donations to help the needy in Germany and Austria. Four other Northwest states and Alaska also supported the drive. The ship's cargo included Tacoma-made wheat and raisins donated by Brown & Haley. (Northwest Room, Tacoma Public Library, Richards Studio D31695-3.)

In 1947, Kaiser Aluminum, under the name Permanente Metals, leased the former Olin Corporation plant in the Tideflats. The plant had been built by the Defense Plant Corporation to produce aluminum for the war effort. Closed after the war, the plant reopened in 1947. Some of its aluminum was used in new products, such as aluminum safety shoes (pictured). (Northwest Room, Tacoma Public Library, Richards Studio D31250-1.)

On October 27, 1948, aluminum plant employees and their families gathered to celebrate the first anniversary of Permanente Metals. The older girl in the photograph is sitting on a stack of aluminum "pigs" (large blocks) that had been produced in the Tideflats plant. (Northwest Room, Tacoma Public Library, Richards Studio D35733-11.)

This 1949 photograph shows aluminum plant workers pouring molten aluminum into a mold to form pigs. Once the pigs were cooled, they were stacked on pallets and sent to the aluminum rolling mill in Trentwood, near Spokane, Washington. The pigs generally weighed between 50 to 1,000 pounds. (Northwest Room, Tacoma Public Library, Richards Studio D46292-1.)

This 1951 photograph shows an employee of the St. Paul & Tacoma Lumber Company working to ensure that lumber gets safely loaded onto a ship. Years earlier, the company's lumber had played an important part in rebuilding San Francisco after its 1906 earthquake. This company became part of the St. Regis Company in the late 1950s. (Northwest Room, Tacoma Public Library, Richards Studio A60342-2.)

This July 2, 1953, aerial photograph shows the vast amount of vacant land the port was working to develop. Mothballed naval vessels are in the foreground of the photograph, and the Port Industrial Waterway Bridge (later Blair Bridge) is shown under construction. The bascule bridge would provide ship access to the terminals and facilities that were later developed along the waterway. (Northwest Room, Tacoma Public Library, Richards Studio D75943-16.)

In 1950, the Purex Corporation bought five acres of land from the port to build a bleach plant at Lincoln Avenue and Thorne Road. The land was part of the port's 300-acre area known as the Port Quadrangle. This photograph shows a crane lifting a tank into place at the Purex plant. (Northwest Room, Tacoma Public Library, Richards Studio D46417-2.)

The Stauffer Chemical Company opened a plant in the port area on May 8, 1950. This group photograph shows company leaders at the plant's grand opening event. The company manufactured agricultural chemicals, including soil sterilants, insecticides, and soil conditioners. The plant produced about 30,000 tons of products in its first year. It was located at 2545 Lincoln Avenue. (Northwest Room, Tacoma Public Library Richards Studio D49856-36.)

On May 23, 1950, a worker was putting the finishing touches on a large tank at the Barthel Chemical Construction Company. The company worked with brick, tile, and rubber pipe linings to create acid and alkali-proof construction. Barthel worked on planes for Boeing and tanks for the Navy. The company was located at 2434 East Eleventh Street. (Northwest Room, Tacoma Public Library, Richards Studio D50225-4.)

In 1950, voters passed a $1 million bond issue for the construction of the Port Industrial Waterway Bridge. Completed in 1953, the bascule bridge had a 150-foot opening so ships could access the upper waterway. The bridge also reduced Tideflats commuting time. The bridge and waterway were later renamed Blair in honor of port commissioner Archie Blair. (Northwest Room, Tacoma Public Library, Richards Studio A80571-6.)

On July 22, 1995, comedian Chico Marx (center) entertained the crowd at the groundbreaking ceremony for US Oil and Refining Company's plant in the Tideflats. Tacoma mayor Harold Tollefson stands to the left of Chico. The company built a $10 million facility on 120 acres of land it bought from the port in the Port Quadrangle. (Northwest Room, Tacoma Public Library, Richards Studio D92109-6.)

In 1951, Concrete Engineering Company (now Concrete Technology) bought port property in the Quadrangle. A leader in prestressed concrete manufacturing, the company worked with Ben Cheney on the design and construction of Cheney Stadium (artist's concept pictured). The stadium opened in 1960 and featured 1,500 concrete pieces. The company has also made concrete piles and deck panels for many port piers. (Northwest Room, Tacoma Public Library, Richards Studio C123400-2.)

This 1948 photograph shows Engine No. 903 pulling freight cars in the Tideflats. The engine was owned by the Tacoma Municipal Beltline Railway (now Tacoma Rail). The line provides important short-line rail service to the port and many of the industries and manufacturing companies located throughout the Tideflats. (Northwest Room, Tacoma Public Library, Chapin Bowen Collection TPL-6892.)

On June 28, 1947, crowds gathered with Chicago-Milwaukee Railroad officials in downtown Tacoma for the christening of the Olympian Hiawatha. The train, which made its inaugural trip from Tacoma to Chicago the next day, cut the travel time from 59 hours to 45 hours. (Northwest Room, Tacoma Public Library, Richards Studio D28582-7.)

The Milwaukee Road (Chicago, Milwaukee, St. Paul & Pacific Railroad Company) opened a new railroad passenger depot in the Tideflats in April 1954. The modern depot, which cost $150,000 to build, was located at 1102 Milwaukee Way. The railroad's popular Hiawatha service, with its diesel-electric locomotive, left from this depot. The Hiawatha made its final run in May 1961. (Northwest Room, Tacoma Public Library, Richards Studio D81872-47.)

This November 9, 1955, photograph shows the model railroad set used by Brakeman Bill on *The Brakeman Bill Show*, a popular local children's television show. In the photograph, Brakeman Bill (Bill McLain) was joined by Phillip McCaffery (left), a railroad car safety inspector. The show left the air in 1974, after more than 20 years on KTNT-TV. (Northwest Room, Tacoma Public Library, Richards Studio D94438-1.)

On December 27, 1954, Johnny's Dock Restaurant opened at the end of Pier 3 at the port. Its unique waterfront location offered diners great views of the nearby fishing boats, port industries, and naval vessels across the waterway. The restaurant, shown in the far lower left of the photograph, was owned by Johnny Meaker. Marshall Perrow was the architect. (Northwest Room, Tacoma Public Library, Richards Studio D88561-2.)

The Top of the Ocean Restaurant, which opened in December 1946, was located on Tacoma's waterfront at 2217 Ruston Way. Constructed to resemble a ship, it was built on pilings at the cost of $262,000. It was called "the most beautiful luxury dine and dance liner." The restaurant was destroyed in an arson fire on April 3, 1977, and never rebuilt. (Northwest Room, Tacoma Public Library, Richards Studio D24966-3.)

In the early 1950s, the Oerlikon Engineering Company of Zurich, Switzerland, bought part of the old Henry Mill site. The company manufactured industrial-grade circuit breakers. It selected Tacoma because of the abundance of skilled labor in the area, especially machine tool operators. The facility was located at 3001 North Starr Street. (Northwest Room, Tacoma Public Library, Richards Studio D67631-2.)

Six

Bold Expansion Plans and Investments

This 1954 aerial photograph shows mothballed vessels at the Tacoma Naval Station on the left, and lumber, grain, and logging activities on the right. In 1959, the port bought the 182-acre station from the federal government for $2.1 million. Originally called the Port Industrial Yard, the port later renamed it the Earley Business Center to honor Robert Earley, a port commissioner. (Northwest Room, Tacoma Public Library, Richards Studio D36257-5.)

In 1955, the port hired Tippetts, Abbett, McCarthy, Stratton (TAMS), a New York planning firm, to develop a long-range plan for port growth. Their report recommended major investments to dredge and extend the port's two major waterways. This August 1963 aerial photograph shows dredge work on the Hylebos Waterway as well as land that had been filled in with dredge material. (Northwest Room, Tacoma Public Library, Richards Studio D139204-18.)

A $5.4 million bond issue helped provide the port with funds to implement the TAMS plan, which created about four more miles of waterfront space for terminals and about 1,500 acres of land for industrial use. This 1974 aerial photograph shows a wide range of industrial and shipping activity taking place on the expanded and improved Hylebos Waterway. (Northwest Room, Tacoma Public Library, Richards Studio D165022-13C.)

The Blair Waterway dredging was completed in 1966, making the waterway 2.6 miles long. The dredge material was used to reclaim adjacent marshy land. Three new shipping facilities soon sprang up along the waterway: Blair Terminal (1972), Weyerhaeuser Wood Chip Facility (1973), and Pierce County Terminal (1973). Two warehouses were also added to the Pierce County Terminal in 1974. (Northwest Room, Tacoma Public Library, Richards Studio D147400-18.)

This 1995 aerial photograph shows a wide range of cargo and shipping activity on the Blair Waterway. Log facilities can be seen on the left. Although logs were once a major part of the port's business, the last log ship called at the port in 2020. Many of the facilities shown in this photograph were later redeveloped into container terminals, auto facilities, and warehouse/distribution centers. (Port of Tacoma.)

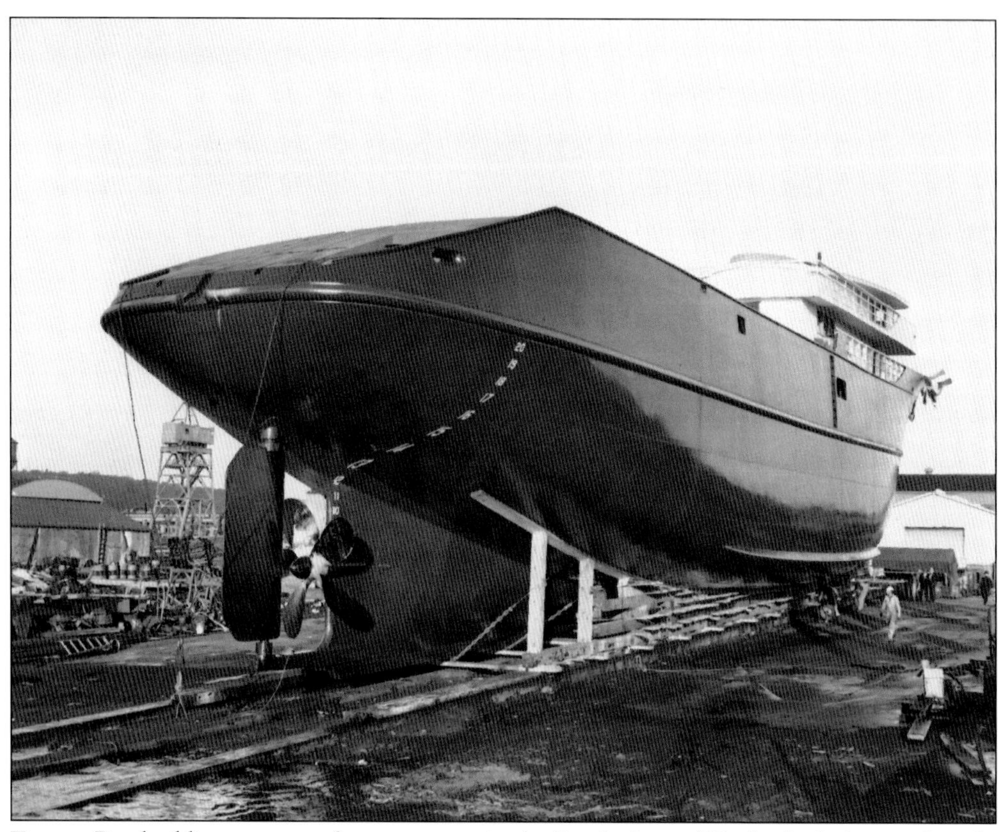

Tacoma Boatbuilding was one of many tenants in the Port Industrial Yard, which the port bought from the federal government. A center for shipbuilding in World War II, the company carried on the shipbuilding tradition there. It manufactured nine tuna seiners between 1966 and 1970. The 191-foot-long *Captain Vincent Gann* (pictured) was launched on March 22, 1970. (Northwest Room, Tacoma Public Library, Richards Studio D158417-19.)

Star Iron and Steel Company was also a tenant in the Port Industrial Yard. This 1970 photograph shows the crew that was building 100-ton floating cranes for the US Navy. Back in 1950, the company had also built smaller, 50-ton capacity cranes for the Navy. (Northwest Room, Tacoma Public Library, Richards Studio D159026-5.)

In the 1960s and 1970s, the port received $7.2 million from the Economic Development Administration (EDA), which helped fund many key projects, including a warehouse at Terminal 7 and two warehouses (pictured) near the Blair Waterway. Other EDA funds helped the port develop Terminal 4 on the Blair Waterway. (Northwest Room, Tacoma Public Library, Richards Studio D166700-84C.)

In 1974, the port leased 101 acres on the Blair Waterway to an Alaskan pipeline contractor building huge oil field modules for Alaska's North Slope. The modules were loaded onto barges in a port-built barge slip and sent on a 4,000-mile journey to Prudhoe Bay. A new "crop" of modules was built there each year through the 1980s, creating thousands of construction jobs. (Port of Tacoma.)

The "Original Tacoma Dome" is shown in this 1967 aerial photograph. The dome, located at Terminal 7, stored alumina imported from Australia. The alumina was shipped by rail to the nearby Kaiser Aluminum plant, where it was used to make aluminum. The image also shows ships loading logs as well as log booms in the Sitcum Waterway. (Northwest Room, Tacoma Public Library, Richards Studio D150900-367.)

In 1968, the port established the Frederickson Industrial Development District and bought 510 acres of land 13 miles south of port terminals. Zoned for heavy industry, Frederickson was an ideal location for companies that did not need a waterfront site in the port area. Ostermann & Scheiwe and Spanaway Lumber were two of the first companies to buy land at Frederickson. (Port of Tacoma.)

On November 15, 1968, the Tacoma Propeller Club held a party to celebrate the port's 50th anniversary. Propeller Club president Donald C. Smith (left) congratulated Richard Dale Smith (right), port commission president. Smith later became the port's executive director. The framed artwork in the photograph, highlighting port activities, was presented to the port by the National Bank of Washington. (Northwest Room, Tacoma Public Library, Richards Studio D154000-290R.)

To celebrate "fifty years of progress," the port commissioned the production of commemorative glass dishes. The artwork, which featured port facilities and Mount Rainier in the background, was created by George Jacobson. The dishes proved to be a very popular port promotional item. The artwork was updated every few years, and new versions of the dish were made. They were last produced in 1983. (Photograph by Jessie Koon.)

The port completed a new $19 million grain terminal in 1975. Located on Schuster Parkway, the port leased the facility to Continental Grain Company. The facility handled 432,846 tons of grain during its first six months of operation. Today, the facility is called the Tacoma Export Marketing Company (TEMCO). Corn and soybeans are the two major grains handled there. (Port of Tacoma.)

The Port of Tacoma Administration Building, located on Sitcum Waterway, was officially dedicated on September 22, 1982. The two-story building cost $2.8 million. The building improved efficiency and productivity by centralizing eight port departments that had been previously scattered throughout the Tideflats area. (Port of Tacoma.)

Seven

An Ever-Changing Port

This aerial photograph shows the $5 million, 28-acre terminal the port built for Totem Ocean Trailer Express (TOTE) in 1979. The port later redeveloped this area into a $32 million container terminal for Sea-Land and built a new terminal for TOTE on the Blair Waterway. In the foreground are 41 acres of land the port bought from the Milwaukee Road in 1974. (Port of Tacoma.)

This photograph, taken October 17, 1984, shows two Hitachi container cranes being loaded onto a ship in Kudamatsu, Japan. After a 5,000-mile trip to Tacoma, the cranes were unloaded at the 76-acre container terminal the port was building for Sea-Land on Sitcum Waterway. At the time, Sea-Land was the world's largest container shipping line. (Port of Tacoma.)

In 1985, the top two container shipping lines in the world started calling at the Port of Tacoma. This aerial photograph of Sitcum Waterway shows a Maersk Line ship (left) and two Sea-Land vessels (right). These two lines helped the port's container volumes grow by 236 percent in 1985, signaling Tacoma's emergence as a world-class container port. (Port of Tacoma.)

This aerial photograph shows the Gog-le-hi-te Wetland, the port's first environmental mitigation project. It was built adjacent to the Puyallup River as part of the port's terminal construction project for Sea-Land. By 2022, the port had spent more than $200 million on mitigation and habitat restoration projects that are helping restore and protect habitat for salmon and wildlife. (Port of Tacoma.)

Over time, some port facilities became obsolete. This August 1988 photograph shows the demolition of United Grain, which operated for more than 50 years. The port used the site of the former grain terminal to expand the rail tracks in the North Intermodal Yard in order to meet the growth needs of its container shipping line customers. (Port of Tacoma.)

Opened in 1953, the Blair Bridge gave ships access to waterfront facilities on the upper Blair Waterway. But as ships got larger, some had trouble passing through the bridge safely. The Puyallup Tribal Land Claims Settlement included some funds to resolve the bridge issue. The removal of the bridge in 1997 unlocked the development potential of port and tribal lands on the upper Blair Waterway. (Port of Tacoma.)

In 1981, the port opened the North Intermodal Yard, the first on-dock rail facility on the West Coast. The facility allowed containers to be transferred between ship and rail more quickly and cost-effectively. This photograph, taken in the North Intermodal Yard on March 20, 1991, marked the ceremonial one-millionth container lift in the port's intermodal yards. (Port of Tacoma.)

When Evergreen Line moved from Seattle to Tacoma in 1991, the port's container volumes topped one million for the first time. Their vessels first called at Terminal 4, but as Evergreen's container volumes increased, they outgrew that facility. Evergreen moved to Pierce County Terminal in 2004. This photograph shows an Evergreen vessel docking at that facility at the end of the Blair Waterway. (Port of Tacoma.)

On October 28, 2022, about 50 port retirees and longtime former employees came to the Port of Tacoma Administration Building for a luncheon reunion. Some also took a tour and saw how much the port had changed in recent years. It was estimated that the combined total time the reunion attendees had worked at the port was more than 1,100 years. (Port of Tacoma.)

In the 1980s, the average container ship calling at the Port of Tacoma carried about 2,500 containers or TEUs (Twenty-foot Equivalent Units). Today's largest container ships can handle 24,000 TEUs. This photograph, taken in February 2018, shows a ship bringing four huge new container cranes into the port. The cranes are 295 feet tall and can handle the largest container ships in service today. (Port of Tacoma.)

Eight

Community Connections

The Port of Kitakyushu, Japan, and the Port of Tacoma have been sister ports since 1984. The steel monument in this photograph was part of a gift exchange between the two ports in 1989. It is located along Tacoma's waterfront. Tacoma's other sister ports include Belawan, Kaohsiung, Tianjin, and Vladivostok. The port is also a partner port of the Port of Anchorage. (Port of Tacoma.)

Since 1980, the port has offered free public boat tours once a year. The tours are the port's biggest annual community outreach event, often attracting well over 1,000 people. The narrated hour-long tours give people a unique opportunity to travel on Commencement Bay and watch port activities up close. The port also offers free bus tours numerous times a year. (Port of Tacoma.)

The port built a public observation tower in 1988, which offers close-up views of shipping activity on Sitcum Waterway. The port dedicated it to "the citizens of Pierce County, in recognition of their initial support which led to the official founding of the Port of Tacoma on November 5, 1918, and to their ongoing support which has ensured its continued growth and success." (Port of Tacoma.)

A Tall Ships Festival was held in Tacoma in 2005 and 2008, bringing many historical ships from around the world into Commencement Bay. The festivals also attracted thousands of people who came to see the ships. The City of Tacoma, hundreds of community volunteers, and the port helped sponsor these major community events, giving visitors a greater appreciation of Tacoma's rich maritime heritage. (Port of Tacoma.)

To mark the port's centennial in 2018, the commissioners wanted to create a "legacy project" to last far beyond the centennial year. The port worked in partnership with the City of Tacoma to develop the legacy project. The artwork, entitled *Swell*, was designed by Rotator Creative. It is located at Firefighters' Park at 909 A Street in downtown Tacoma. (Port of Tacoma.)

A spinning globe is part of the international trade kiosk that is on permanent display at the Washington State History Museum in downtown Tacoma. The interactive exhibit highlights various cargoes that move through Washington state's ports and some of the state's major cargoes and trading partners. The port provided funding for the kiosk, which is part of the museum's 360 gallery. (Photograph by Jessie Koon.)

Bibliography

Hunt, Herbert. *Tacoma, Its History and Its Builders: A Half Century of Activity (Vol. 1, 2, and 3).* Tacoma, WA: Tacoma Historical Society Press, 2005. First published 1916 by the S.J. Clarke Publishing Company.

Levinson, Marc. *The Box: How the Shipping Container Made the World Smaller and the World Economy Bigger.* Princeton: Princeton University Press, 2008.

Magden, Ronald, and A.D. Martinson. *The Working Waterfront: The Story of Tacoma's Ships and Men.* Tacoma, WA: International Longshoremen's and Warehousemen's Union, Local 23 of Tacoma, 1982.

———. *The Working Longshoreman.* Tacoma, WA: International Longshoremen's and Warehousemen's Union, Local 23 of Tacoma, 1991.

Mason, Evette, Sherry Maiura, and Bethany Maines. *Port of Tacoma: Gateway to the World: The Story of Pierce County's Port.* Tacoma, WA: Port of Tacoma, 2002.

Oldham, Kit. *Public Ports in Washington: The First Century, 1911–2011.* Seattle, WA: Washington Public Ports Association in association with HistoryLink, 2011.